Engineering Calcula using Microsoft Excel

Learn how to write your own customized calculations in minutes

Primoz Kvaternik, M.Eng.

$i = 0 \rightarrow 6$

$$r^i_{1,M,i} = \frac{M^i_{Ed}}{E_{c,eff} I_1}$$

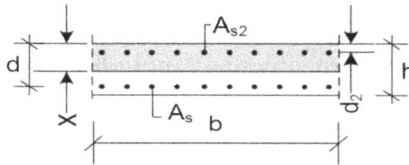

$$r_{1,CS,i} = M^i_{Ed}\, \varepsilon_{cs}\, \alpha_E \frac{S_1}{I_1} \qquad r_{2,CS,i} = M^i_{Ed}\, \varepsilon_{cs}\, \alpha_E \frac{S_2}{I_2}$$

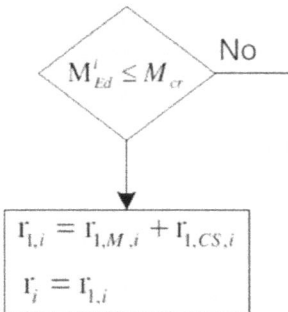

$$M^i_{Ed} \leq M_{cr} \quad \text{No}$$

$$r_{1,i} = r_{1,M,i} + r_{1,CS,i}$$

$$r_i = r_{1,i}$$

T_i	M_{Ed} [kNm]	M' [kNm]	$(1/r)_{i,M}$ [1/m]	$(1/r)_{i,CS}$ [1/m]	
0	0,00	0,00	0,0000	0,0000	0
1	6,50	0,42	0,0024	0,0077	0
2	8,50	0,83	0,0031	0,0100	0
3	9,80	1,25	0,0036	0,0115	0
4	8,50	0,83	0,0031	0,0100	0
5	6,50	0,42	0,0024	0,0077	0
6	0,00	0,00	0,0000	0,0000	0

Deflection calculation

$f_{act} =$ 74,39 mm   Final deflection

GRSOFT Software Development

GrSoft

Published by Primoz Kvaternik, GRSOFT Structural Engineering
http://www.grsoft.eu

First published 2014

ISBN: 978-961-93681-1-4

You can contact us for consultancy at:
http://www.grsoft.eu/   engs@grsoft.eu

Udemy Course
Engineering Calculations using Microsoft Excel at
https://www.udemy.com/engineering-calculations-using-microsoft-excel/#/

Contents

PREFACE

As every Engineer needs to do many daily calculations especially using modern standards like EUROCODES, the need to write custom software solutions is more and more real. Especially if standards include many complex formulas which are hardly calculated using pocket computers as it was 30 years ago. Then it came programmable pocket computers, I clearly remember as I had SHARP programmable computer, where it was possible to write a complex software, but you couldn't print the results as it is possible now. So today it is possible **just by using Microsoft Excel** and its programming abilities to write real software which can solve all daily engineering calculations with ease.

What does an engineer need?

So what does an engineer need when creating calculations? First there are input parameters, which should be entered on a very simple and a quick way, then a simple sketch as a graphical representation of the basis of calculation with annotations of input parameters. After that engineer needs to define the mathematical procedure which could be very simple, but it should also enable him, to write also more complex formulas or iterations. This is very easy to do with Excel.

In this book I will show you that **you do not need to be a software developer to create your own customized engineering calculations in minutes**. What is maybe the most important, you can update formulas in your calculation any time you want. This is the solution that every engineer needs, because it offers open-source solution with powerful programmable tools, but on the other side **simple enough to be done instantly**.

We will learn the following topics:
- How to create cells where input parameters should be entered
- How to create a sketch with annotations of input parameters
- How to prepare cells where results of calculation will be written
- How to create a push button, where you will trigger start of the calculation
- How to write code to perform calculation
- How to write code to display the results of calculation
- How to perform calculation

This book will also show you how to write the software for **practical engineering calculation** for structural analysis. I will show you in detail, how to enter data, define formulas and actually perform calculation, including how to display results and format cells for results of calculation.

I will provide you with an **easy-to-follow material** explanation, all steps including source code will be explained in detail.

Primoz Kvaternik, M.Eng.

ABOUT THE AUTHOR

Primoz Kvaternik, M.Eng.

Primoz is a director of **GRSOFT Structural Engineering Company** in Slovenia / EU. As a Structural Engineer with more than 15 years' experience in structural design on numerous architectural-engineering building projects both large and small, he has been working on project documentation for all types of building structures, reinforcement plans for RC structures, fabrication plans of steel structures, fire resistance calculation of building structures including consultation on the most appropriate fire protection. Based on years of experience and participation in post-earthquake reconstruction, he specialize in Earthquake Reconstruction and rehabilitation of buildings as well.

He has university degree in structural engineering and all possible licenses for the design and revision of project documentation for building structures at **Slovenian Chamber of Engineers**.

He is currently teaching how to use Eurocodes as one of the most advanced building design codes on a world scale. He is teaching also **how to use technology for solving daily engineering calculations, including software development** as a support to structural engineering.

1

VISUAL ELEMENTS ON EXCEL SHEET

In this Chapter

- What are input parameters?
- How to prepare cells for input parameters?
- What are results of calculation?
- How to prepare cells for results of calculation?
- What is calculation sketch?
- How to draw a sketch?
- How to insert a sketch into Excel?
- What is push button?
- How we can add it to Excel?

What are input parameters?

First we need to learn what are input parameters?

Each and every numerical calculation even very simple one has first input parameters which are then used in mathematical calculation. Finally we get results of calculation.

Input parameters ⇒ Math. calculation ⇒ Results of calculation

Let's explain input parameters on the following equation. We need to calculate the value of the function f, by using the following input parameters.

$$a = 1 \quad b = 2 \quad c = 5 \quad x = 10$$

a, b, c, x input parameters

We can see the expression for a function f here

$$f = ax^2 + bx + c = 1 * 10^2 + 2 * 10 + 5 = 125$$

When we enter input parameters into this expression and solve it we get the result of calculation.

How to prepare cells for input parameters?

Let's open one empty excel file

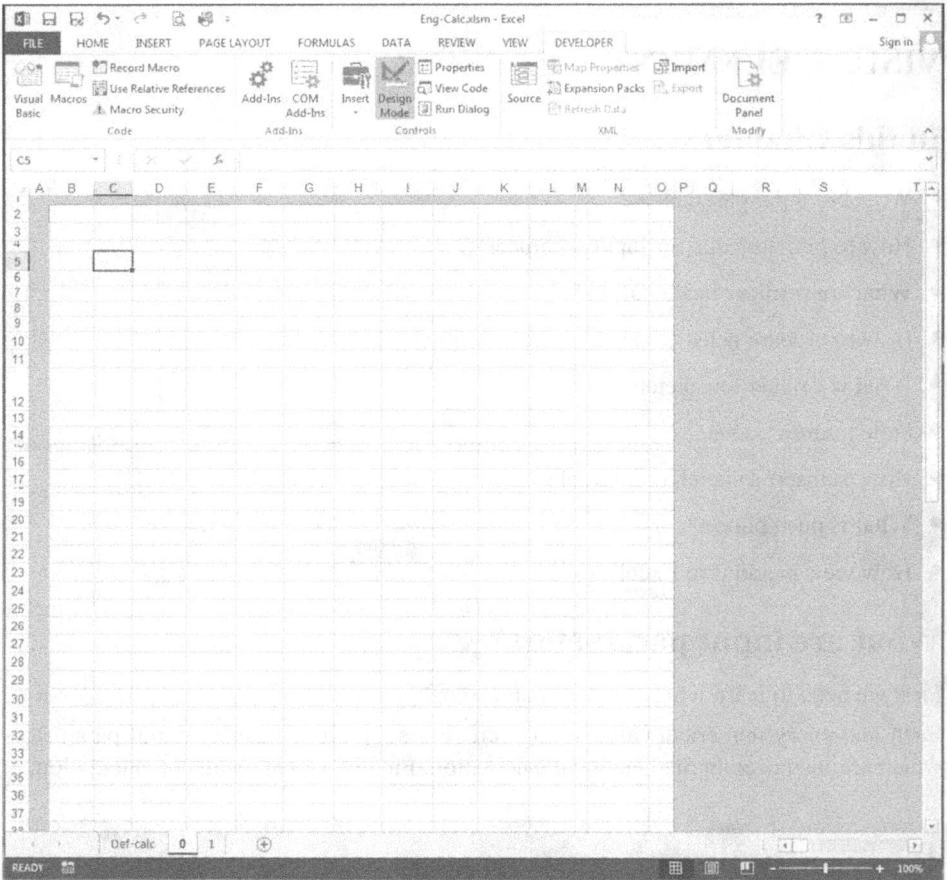

For entering input parameters we need to prepare in Microsoft Excel certain cells where we will type these input parameters which we will later use in mathematical calculation.

Let's create first header for input parameters. We will type here text for the header, then we will put border for it. For this purpose we need first to select these cells with the mouse.

How we can do this?

We click in the first cell and drag the mouse to the other cells containing the text, holding down the left mouse button – when the cells are selected we can release the left mouse button. Now we keep the mouse cursor inside the cells and right click the mouse button.

The popup menu and format toolbar appears

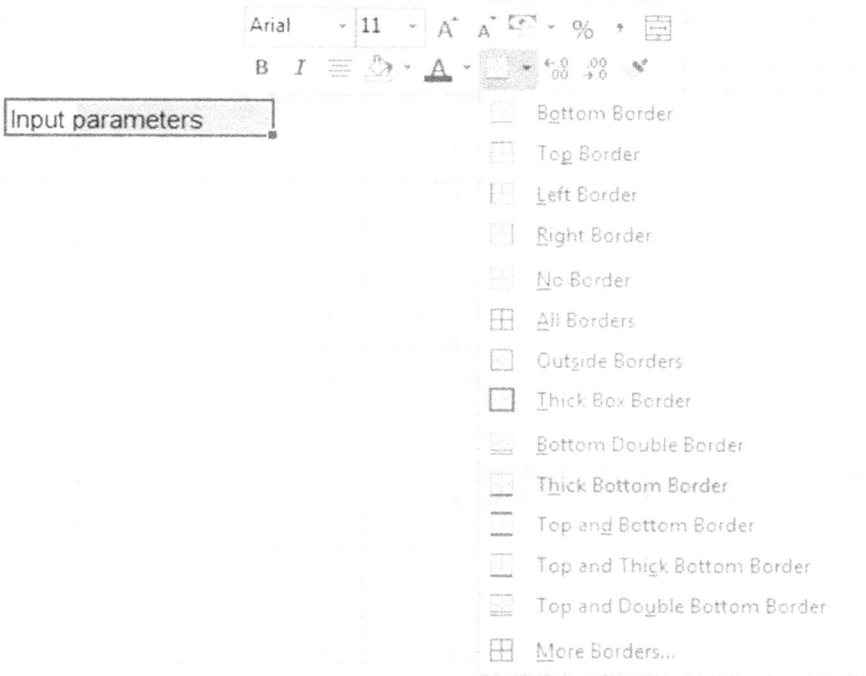

Arial ▾ 11 ▾ A˄ A˅ ✏ ▾ % ▸ ⊞

B I ≡ ◇ ▾ A ▾ ▦ ▾ ←.0 .00 ✔
.00 →.0

Input parameters

	Bottom Border
	Top Border
	Left Border
	Right Border
	No Border
⊞	All Borders
	Outside Borders
□	Thick Box Border
	Bottom Double Border
	Thick Bottom Border
	Top and Bottom Border
	Top and Thick Bottom Border
	Top and Double Bottom Border
⊞	More Borders...

Now we need to select the border for these cells, I personally like the border as shown above, but you have a plenty of borders to choose.

After we selected the border for the cells, we can select a background for this header. For this purpose we select the cells and right click the selection on the same way as we did before and choose from the toolbar appropriate fill color as shown on the next page

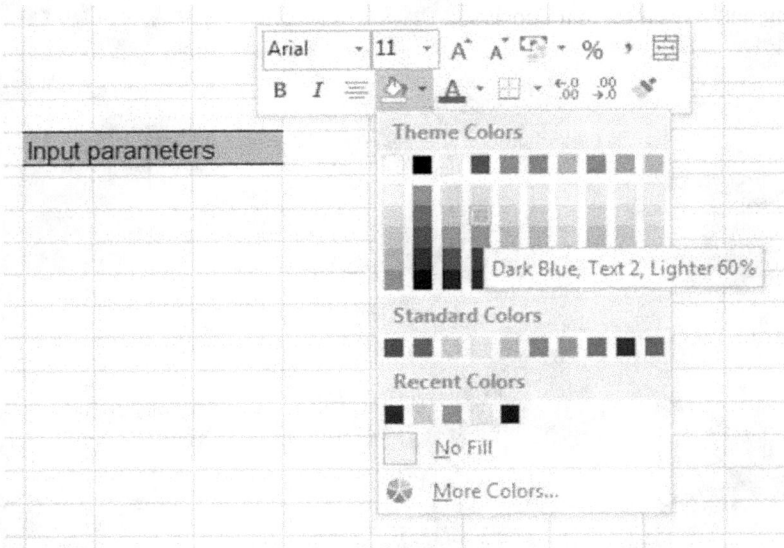

After we created the header for input parameters, we need to create cells for first input parameter. For this purpose we type the name of input parameter, its value and description as it is shown below.

Input parameters

b=   100,00 cm   Width of the cross section

We want also to format the number value – for this purpose we need to right click over the number cell and choose format cells from the menu

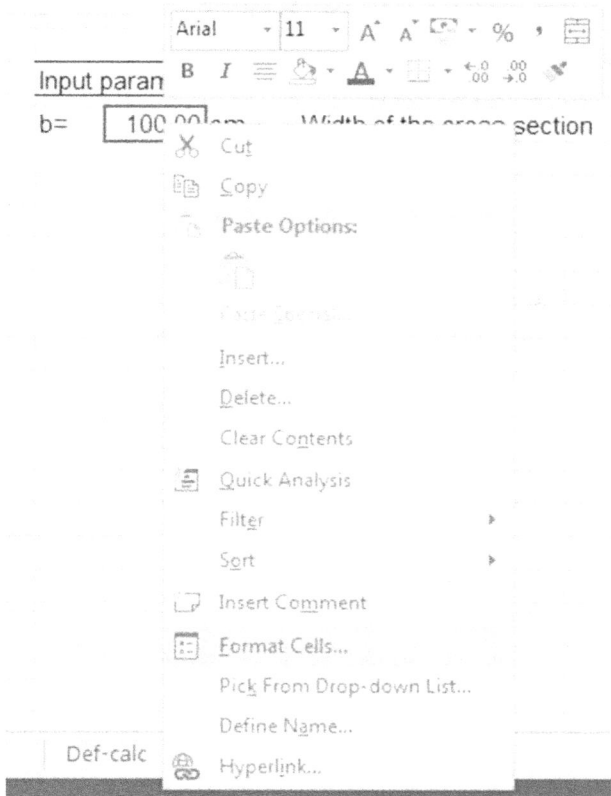

Arial	11	A A	%	
B	I	A		

Input paran

b=   100.00   Width of the cross section

- ✂ Cut
- ▤ Copy
- Paste Options:
  - ▯
  - Paste Special...
- Insert...
- Delete...
- Clear Contents
- ▤ Quick Analysis
- Filter ▸
- Sort ▸
- Insert Comment
- Format Cells...
- Pick From Drop-down List...
- Define Name...
- Hyperlink...

Def-calc

After we choose format cells menu item, we can select the number format and the number of decimal places for the cell as shown on the next page.

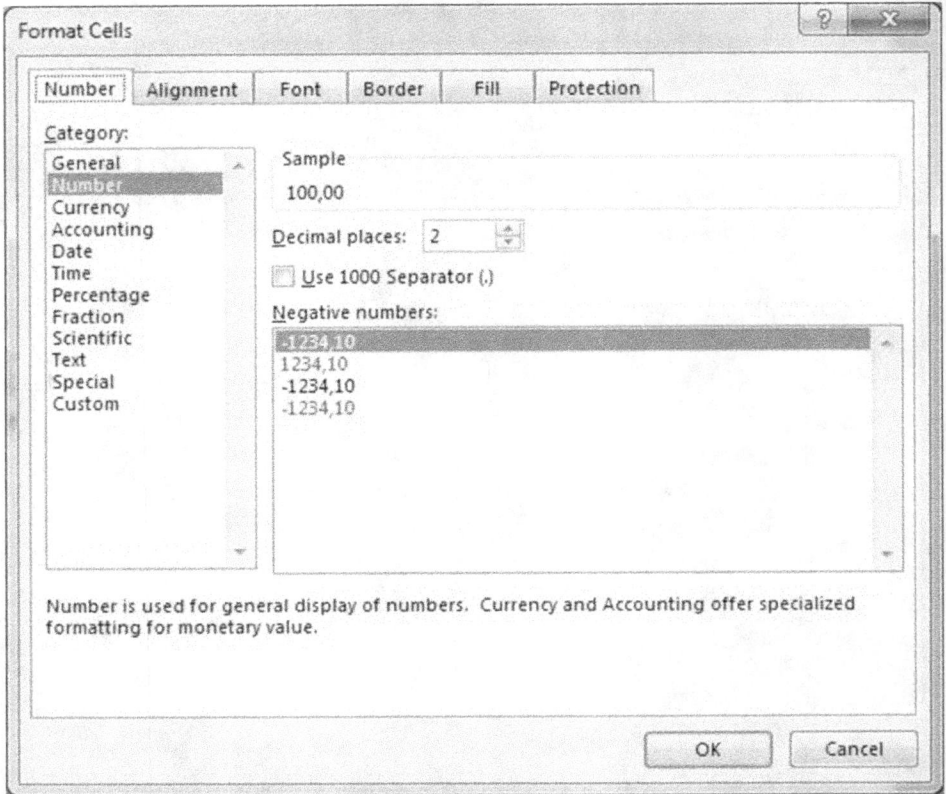

It is recommended that we choose color for font as input parameters should have text color different as calculated values.

On the same way we create cells for all other input parameters which we may have in our mathematical calculation.

What are results of calculation?

As we already said, results of calculation or so called output parameters are the results which Excel calculates in mathematical calculation. They are stored in variables and need to be displayed in Excel worksheet.

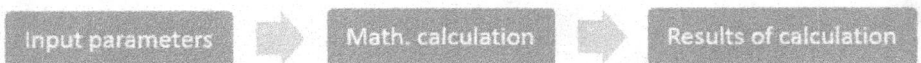

How to prepare cells for results of calculation?

For the results of calculation we need to prepare in Microsoft Excel certain cells where they will be displayed. Let's create first the header for results of calculation. The procedure is almost the same as we have done for input parameters.

First we will type text for the header, then we will put border for this header, for this purpose we need first to select these cells with the mouse. We will do this as we did before. We click in the first cell and drag the mouse to other cells holding down the left mouse button – when the cells are selected we can release the left mouse button.

Now we keep the mouse cursor inside the cells and right click the mouse button. The popup menu and format toolbar appears, where we select the border and the background for these cells on exactly the same way as we did before.

After we created header for results of calculation, we need to create cells for first output parameter. For this purpose we type the name of output parameter, its value and description as it is shown below.

Curvature calculation

$M_\sigma=$ 9,75 kNm   Cracking moment

We want also to format the number value – for this purpose we need to right click over the number cell and choose format cells from the menu. The procedure is exactly the same as we already explained at input parameters.

It is recommended that we choose color for font as input parameters should have text color different as calculated values.

On the same way we create cells for all other output parameters which we may have in our mathematical calculation.

What is calculation sketch?

So what is calculation sketch and why does an engineer need it?

A sketch is graphical representation of calculation as it represents the basis of calculation and its input parameters. So it is evidently a good praxis to show it at the beginning of the calculation sheet.

How to draw a sketch?

We will use Microsoft Visio 2013, which is part of Microsoft office 2013, but you can have also older version of office. This software is very good for drawing sketches which you can easily import into the Excel.

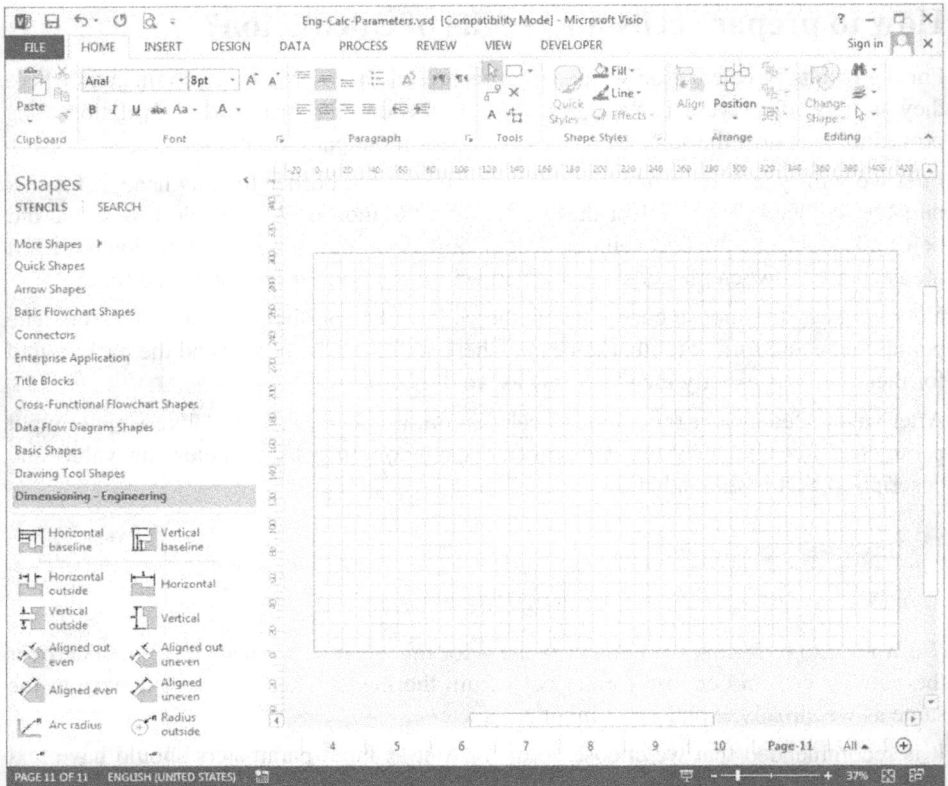

In this example we will draw a cross section of a reinforced concrete slab with annotations of input parameters. We will not go here into details how to use Microsoft Visio, as this is not the topic of this book, but rather just to show an example of how to quickly draw a sketch which is to my opinion the easiest way to have it in Excel. You can use any other tool as well and insert the sketch into the Excel on the same way as I will show you later.

First we will draw an outline of the slab. For this purpose we select the line tool from the main toolbar.

REVIEW   VIEW   DEVELOPER

Pointer Tool   Fill ·
Connector   Rectangle   Line ·
A Text   Ellipse   Effects·
Tools   Line   ies
Freefc   Line (Ctrl+6)
Arc
Pencil

We will draw a line by clicking a start point and then dragging the cursor to the last point and holding down the left mouse button and release it at the end. The line is drawn as shown below

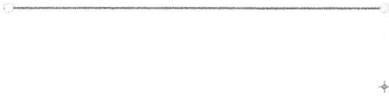

Now we change the line tool to the pointer tool in the main toolbar

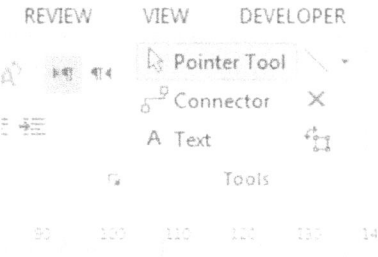

REVIEW   VIEW   DEVELOPER

Pointer Tool
Connector
A Text
Tools

We need to copy this line 2 times. For this purpose we press the ctrl key and drag the line to the new position and then release the ctrl key. We will do it twice as we need two lines as shown below.

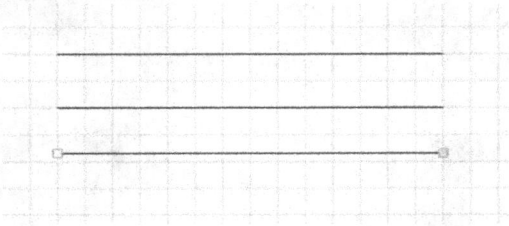

Now we will draw the left vertical border line using the same line tool and copy this line to the right side using the ctrl key again.

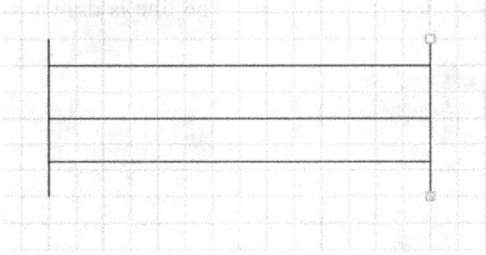

As we want the lines to have greater thickness and vertical lines to be dashed we will first select all lines with the mouse and then click the line shape styles button as shown below.

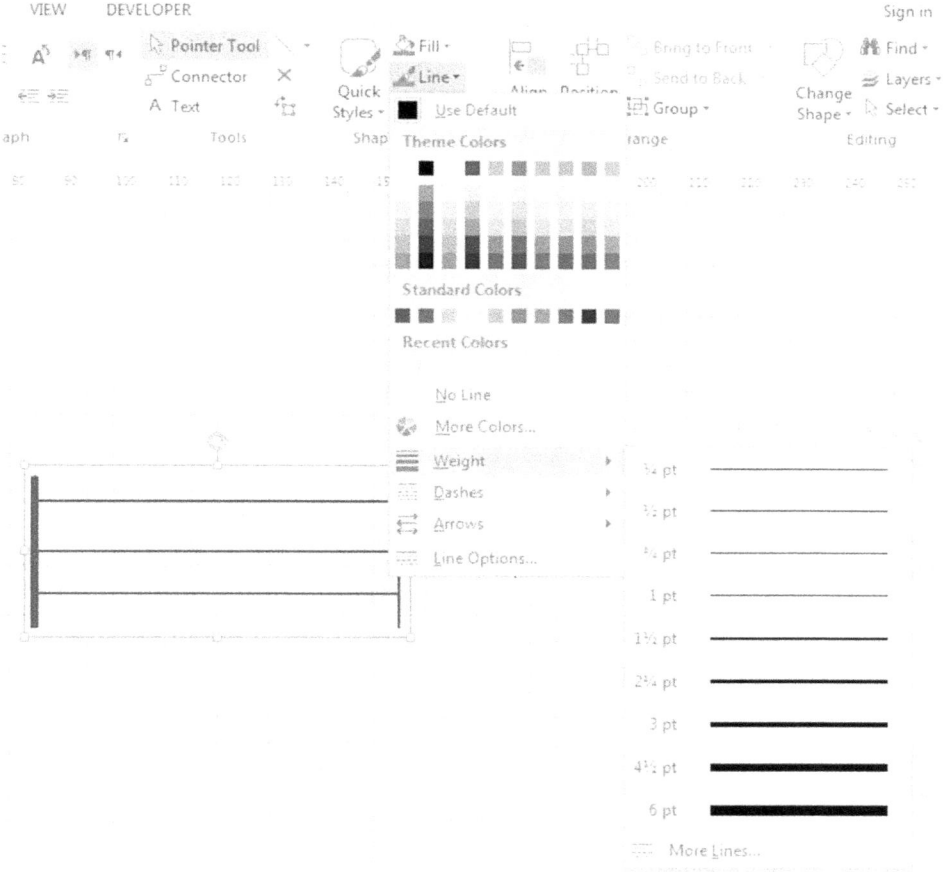

In the popup window above we select the weight for all lines as 1 point. Then we select just vertical border lines with the mouse and define them as dashed on a completely the same way.

Now we need to create reinforcement bars. For this purpose we will select an ellipse tool to draw a circle as shown below.

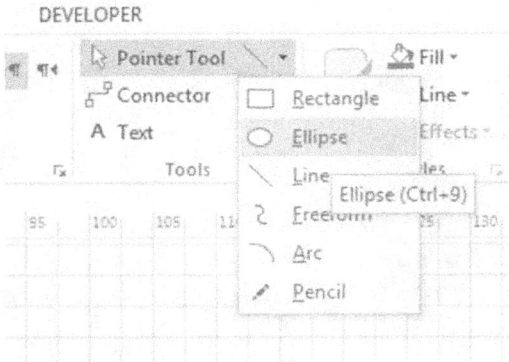

DEVELOPER

- Pointer Tool
- Connector
- A Text
- Rectangle — Line
- Ellipse — Effects
- Line
- Ellipse (Ctrl+9)
- Freeform
- Arc
- Pencil

Tools

When the circle is drawn we fill it with black color using the fill button on the shape styles toolbar as shown below.

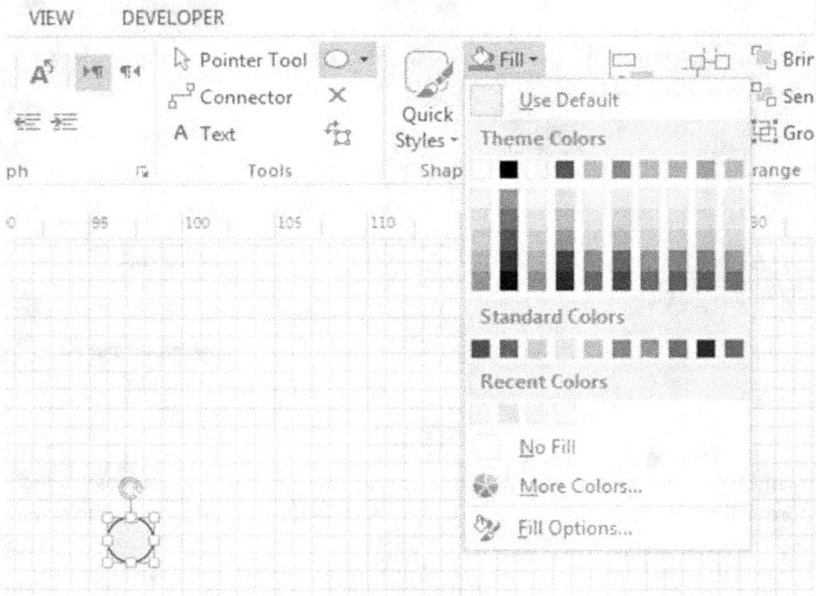

VIEW   DEVELOPER

- Pointer Tool
- Connector
- A Text
- Fill
- Quick Styles
- Use Default
- Theme Colors
- Standard Colors
- Recent Colors
- No Fill
- More Colors...
- Fill Options...

Now we will move this circle to the cross section and we will replicate this circle in the cross section according to our needs so that we will get upper reinforcement of the cross section. On the same way we will copy top line of circles to the bottom part of the cross section as shown below.

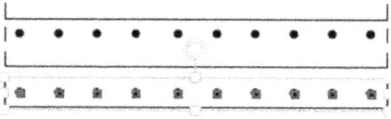

Then we need to fill the upper part of the cross section into gray. For this purpose we will draw rectangle form the main toolbar, fill it with gray color and select no-line from the line menu, to avoid border line of the rectangle.

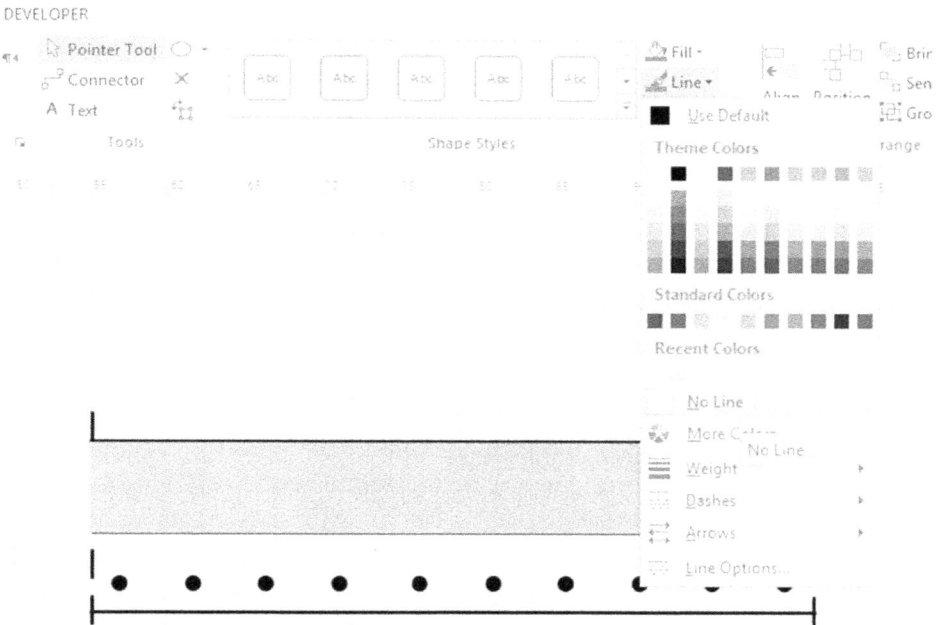

Then we need just to put this rectangle to the back of the other drawing elements. For this purpose we click **Send to back** button from the main toolbar.

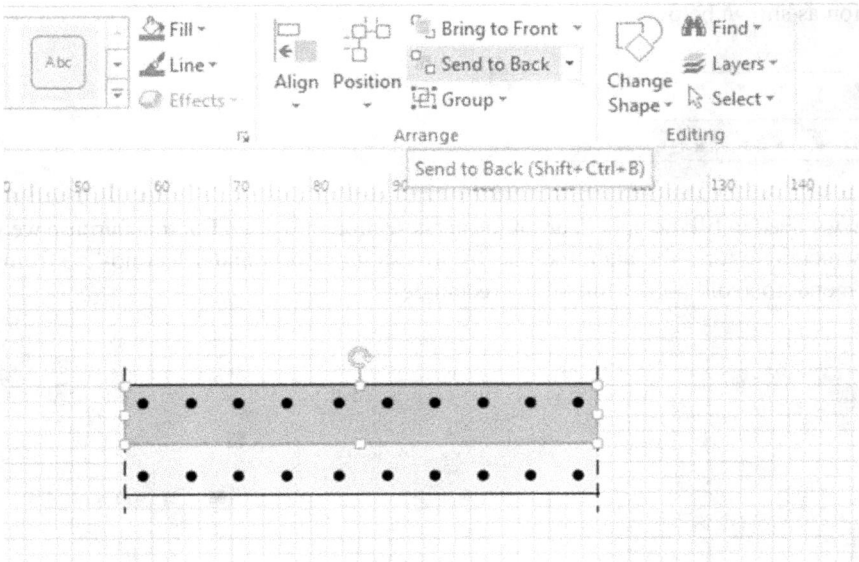

We need also to draw dimension lines to the cross section. We will use first **horizontal baseline** from Dimensioning-Engineering stencils.

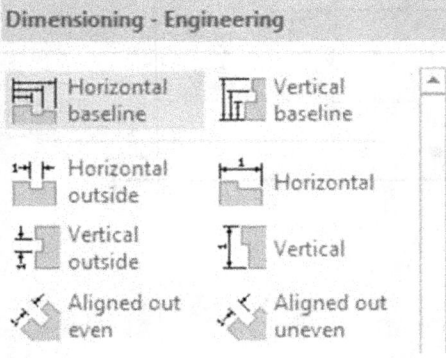

We drag this tool from the toolbar and holding down the left mouse button to the desired place in the drawing and then release left mouse button. We need to resize the dimension line and change the arrow type in the format shape toolbar as seen below.

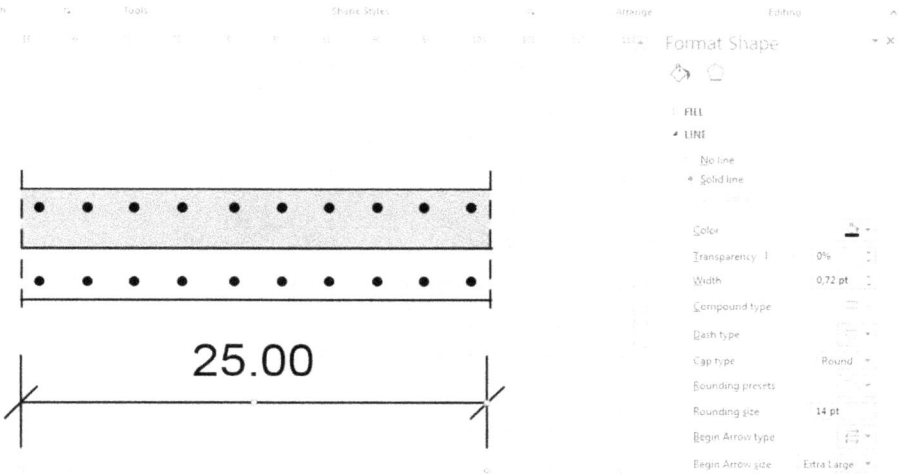

Format Shape   ▾ ✕

FILL

⊿ LINE

   No line

   ● Solid line

Color	
Transparency	0%
Width	0.72 pt
Compound type	
Dash type	
Cap type	Round
Bounding presets	
Rounding size	14 pt
Begin Arrow type	
Begin Arrow size	Extra Large
End Arrow type	
End Arrow size	Extra Large

25.00

To change the text of the dimension line into parameter b we just right click the dimension line and select **Edit text** from the menu. We enter letter b and change the font characteristics from the main toolbar.

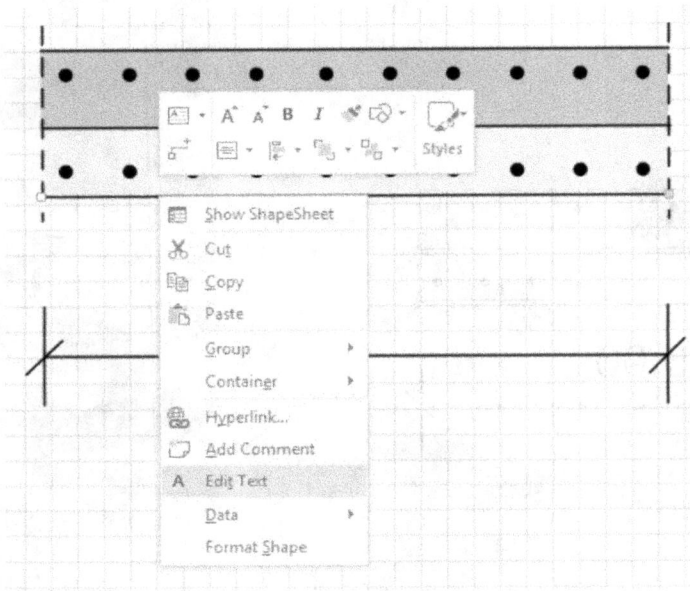

To draw vertical dimension lines we will use the **vertical baseline** from Dimensioning-Engineering stencils on the completely the same way as for horizontal dimension lines.

To draw text annotations we use the text tool from the main toolbar as shown below.

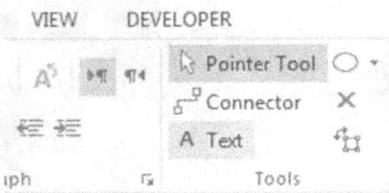

All other elements of the sketch are created on the exactly the same way and at the end we have the following sketch.

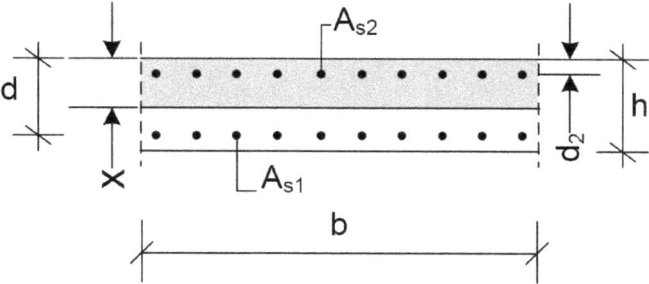

How to insert a sketch into Excel?

When the sketch is finished we select the sketch with the mouse and copy it clicking the copy button on the main toolbar.

Now we leave Visio opened and open Excel file where we will have our calculation.

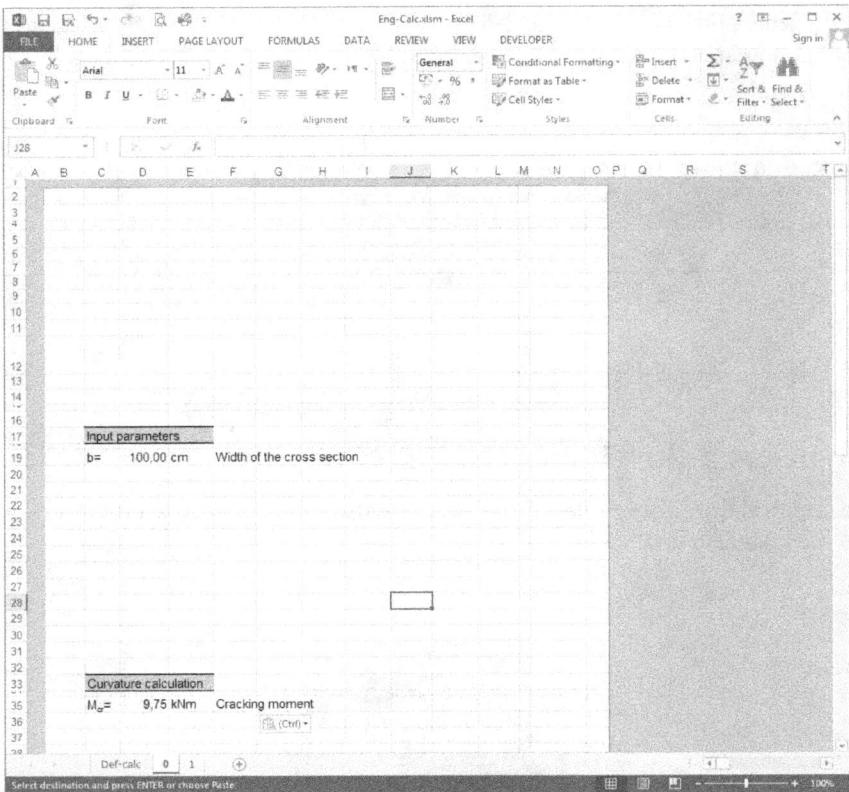

We will paste a sketch into Excel using Paste Special and Enhanced Metafile command as shown below.

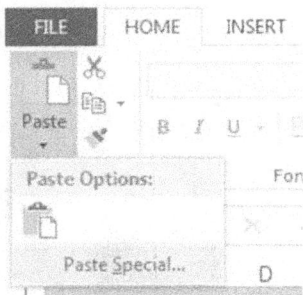

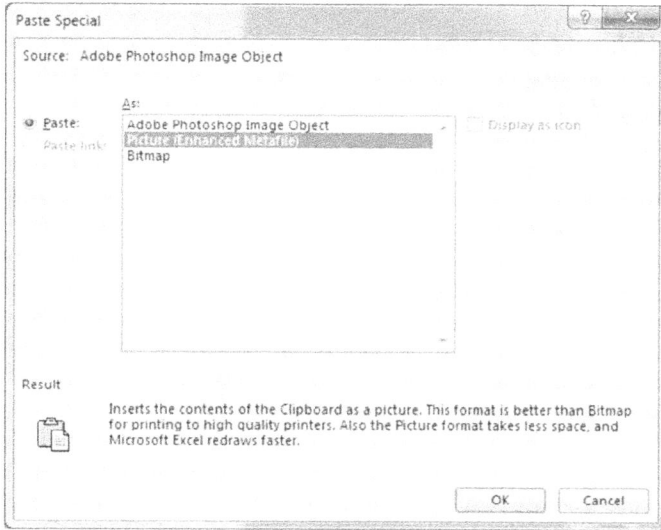

This way we avoid borderline of the imported sketch as we can see below.

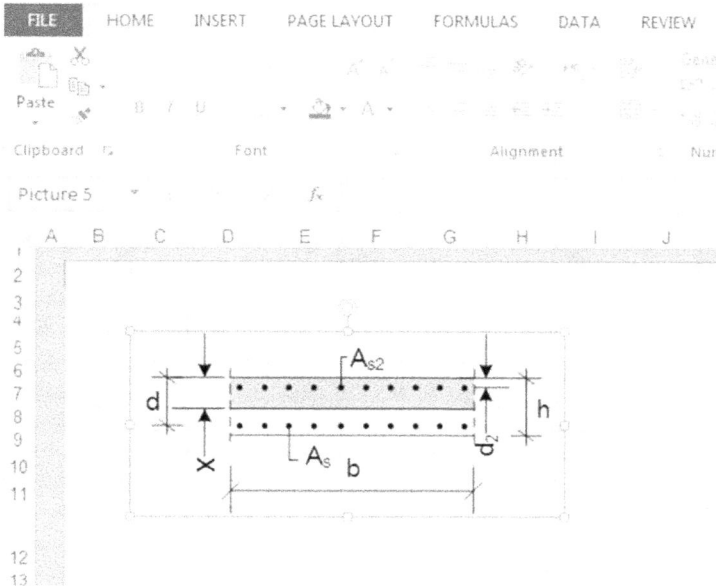

So the sketch is imported into the Excel worksheet, we can just position it on the desired place. Our sketch has all important input parameters annotated and serves as an additional information of our engineering calculation.

You could create a sketch also using any other software, but we used Microsoft Visio which is part of Microsoft Office, because we think it is the easiest way.

What is Push Button?

As this course is **intended** for **complete beginners** who are not involved into software development at all, we will start at the beginning.

So first we need to know what Push Button is.

It is a graphical representation of a button which we put into the excel worksheet. Its purpose is that when we click on it with left mouse button, specific action is performed.

We can see a sample of a push button below.

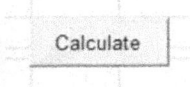

How we can add it to Excel?

So we need to know first how we can add it into the Excel worksheet. First we need to have the **Developer Tab** visible in the Excel Ribbon. For this purpose we need to go to the Excel Options dialog box, by clicking on the File Tab and select Options menu item. The following dialog box opens.

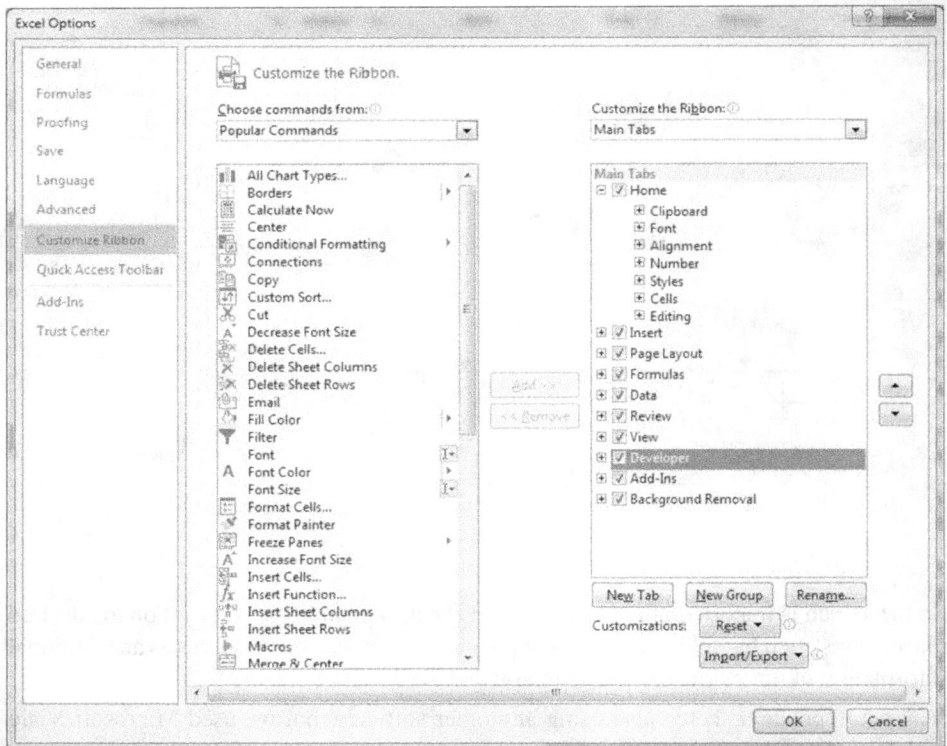

We need to go to Customize Ribbon Tab and check Developer check box in Main Tabs list box. This check box should be checked, that Developer Tools are activated in Excel. We confirm this selection by clicking the OK button and close this dialog box

Now we need to select a Developer Tab on a Ribbon and click the Insert Button. When we do this, we can see a lot of controls available in drop down list offered as we can see below.

We need to select command button from Active X controls by clicking it. When we do this a cross shaped cursor appears and we are expected to draw somewhere in the worksheet a rectangle which represents this button as we can see below.

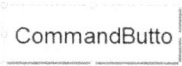

To change the text written in the button, we put mouse cursor inside the button area, right click it and then select properties from the menu which appeared. The properties window appears.

Properties

CommandButton1 CommandButton

Alphabetic | Categorized

(Name)	CommandButton1
Accelerator	
AutoLoad	False
AutoSize	False
BackColor	&H8000000F&
BackStyle	1 - fmBackStyleOpaque
Caption	Calculate
Enabled	True
Font	Arial
ForeColor	&H80000012&
Height	29,25
Left	334,5
Locked	True
MouseIcon	(None)
MousePointer	0 - fmMousePointerDefault
Picture	(None)
PicturePosition	7 - fmPicturePositionAboveCenter
Placement	2
PrintObject	True
Shadow	False
TakeFocusOnClick	True
Top	417,75
Visible	True
Width	85,5
WordWrap	False

Calculate

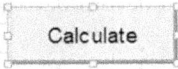

We will first change the name of the button to a new one, then we will change the caption property of the button to Calculate, as this is what this button will do. As we do not want that this button is printed when we print worksheet, we will set **PrintObject** property to false.

So we created the push button, now we can just resize it and place on the desired position on the worksheet.

I should mention here that design mode should be activated when we are changing the properties of the button – this is done by default when we create it, but later you need to click **Design mode** button manually. You can check if you are in design mode, if the button can be selected, otherwise you are in the Run mode – about which we will talk later.

In this chapter we just learned how to visually create the push button.

2

HOW TO WRITE SOFTWARE CODE?

In this Chapter

- Code Editor
- Creating a Sub Procedure
- Creating variables
- Reading input parameters
- Entering formulas

Code Editor

So what is code editor and what is code at all?

These are very fundamental questions especially for the ones who are not involved into software development at all – and this book is intended just for that kind of people. This book will provide you with a very simple answer to this question, with the answer which is clear enough that can be understood just by everyone.

So what is code?

Code is a sequence of instructions which we give to the Excel to do some work for us – instead that we do it manually in Excel, or write formulas in Excel, we write the code which Excel later performs.

What is code editor?

It is the text editor, where we write instructions in the language which Excel understands, as Excel does not speak English, we need to learn the language that Excel speaks – this language is called Visual Basic for Application or shortly VBA.

We can see the code editor below.

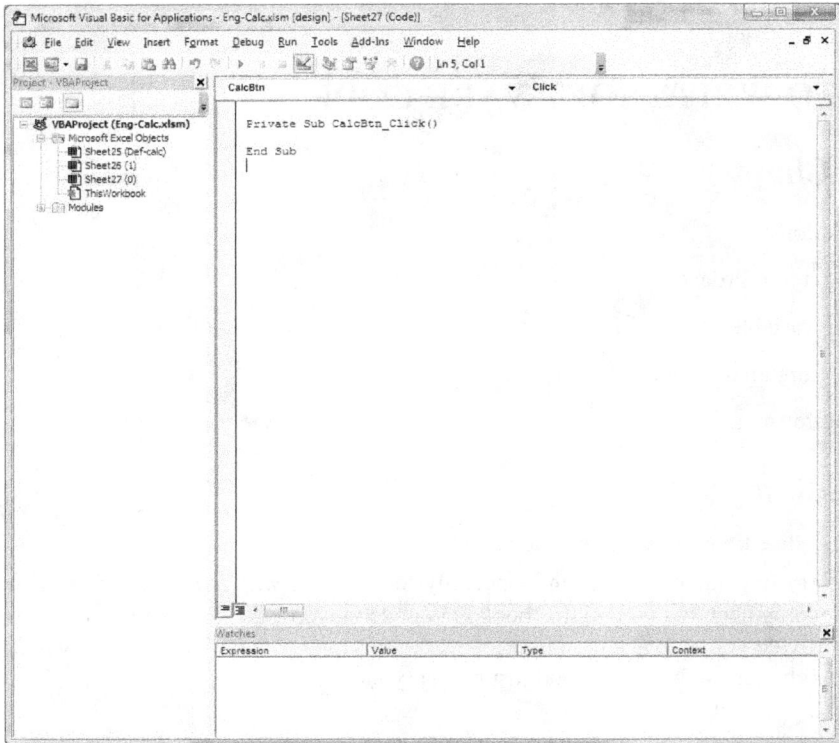

So in code editor we write a letter to Excel in a language which Excel understands...we can say that we write code and this code is part of the Excel worksheet and is saved together with the Excel file.

In code editor we can do many things, but now we will concentrate on the topic we have and will not go further as we want to explain the whole thing simple enough even for complete beginners.

In this chapter we will learn how to write this code in code editor and how to connect it with the data we have in our worksheet.

Creating a Sub Procedure

So what is Sub Procedure and why it is so important?

Sub procedure is a series of Visual Basic for Application statements which are enclosed between Sub and End Sub statements. These statements are executed, when we call Sub Procedure.

The easiest way to create a Sub Procedure for our worksheet is to select first Design Mode from Developer tab on Ribbon, so that we can select the push button. When the push button can be selected as we see below, we just double click on the button.

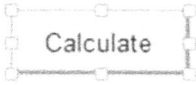

```
    Calculate
```

When we do this, code editor is displayed and default Sub Procedure is created and already connected with our push button.

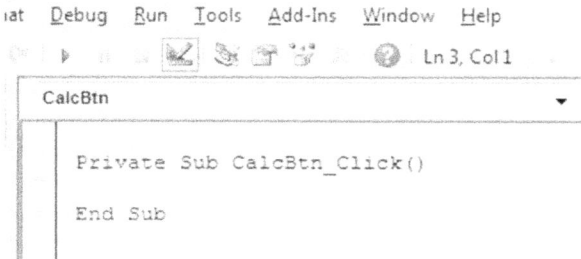

```
ıat   Debug   Run   Tools   Add-Ins   Window   Help
                                      Ln 3, Col 1

  CalcBtn                                      ▾

      Private Sub CalcBtn_Click()

      End Sub
```

Now we need just to write specific statements between Sub and End Sub in the language which Excel understands.

In the following lines we will learn how to write statements here.

Creating variables

Let's first explain what variables in any calculation are.

We will learn this on this simple formula, where we want to calculate the sum of two numbers.

$$b = a + c$$

If we want to store the result of this calculation, we need to define variable **b** for storing this numerical value.

We learned what is variable in mathematics, now we need to know how are variables defined in the language which Excel understands?

In Excel we define **simple variable** which stores numerical value by typing the following statement into the code editor. First we will type comment line following with the numeric variable declaration as seen below.

```
Debug  Run  Tools  Add-Ins  Window  Help
 ▶  ‖  ▪  🔲  🔍  🔚  🔛  ⚙  ❓  Ln 8, Col 1
─────────────────────────────────────────────
CalcBtn

  Private Sub CalcBtn_Click()

  ' Declaring simple variables
  Dim b As Double

  End Sub
```

On exactly the same way we create as much variables as we need in our calculation.

If the result of calculation is not stored just in one variable but instead in a field of variables, we say that it is stored in an **array**.

Let's learn what is an array?

T_i	M_{Ed}	M	$(1/r)_{I,M}$	$(1/r)_{I,cs}$	$(1/r)_{II,M}$	$(1/r)_{II,cs}$
	[kNm]	[kNm]	[1/m]	[1/m]	[1/m]	[1/m]
0	0,00	0,00	0,0000	-0,0012	0,0000	-0,0052
1	6,50	0,42	0,0024	0,0012	0,0064	0,0052
2	8,50	0,83	0,0031	0,0012	0,0083	0,0052
3	9,80	1,25	0,0036	0,0012	0,0096	0,0052
4	8,50	0,83	0,0031	0,0012	0,0083	0,0052
5	6,50	0,42	0,0024	0,0012	0,0064	0,0052
6	0,00	0,00	0,0000	-0,0012	0,0000	-0,0052

If we want to store the values of the third column from upper table, we can say that we have an array named M_c which stores these values from the table. Each value in the array is accessible with an index which runs from 0 to 6.

Now we need to know how we define an array which stores numerical values in Excel.

In Excel we define **array variable** which stores numerical values by typing the following statement into the code editor. First we will type comment line following with the array variable declaration as seen below.

```
Debug   Run   Tools   Add-Ins   Window   Help
  ▶        ☑  ⤳  ⤵  ⤴        ②   Ln 11, Col 1
CalcBtn

    Private Sub CalcBtn_Click()

    ' Declaring simple variables
    Dim b As Double

    ' Declaring arrays
    Dim Mc(6) As Double

    End Sub
```

On exactly the same way we create as much arrays as we need in our calculation.

Reading Input Parameters

Before we start entering formulas, we need to store input parameters into variables, so that we can later use them in our calculation.

We can do this in Excel with the following statement.

```
Debug   Run   Tools   Add-Ins   Window   Help
  ▶        ☑  ⤳  ⤵  ⤴        ②   Ln 14, Col 1
CalcBtn                                          ▼

    Private Sub CalcBtn_Click()

    ' Declaring simple variables
    Dim b As Double

    ' Declaring arrays
    Dim Mc(6) As Double

    ' Reading input parameters
    b = ActiveSheet.Cells(19, 4)

    End Sub
```

It means that we will store the value entered in excel worksheet in the cell located at row 19 and column 4 into variable **b**. So variable **b** will from now on hold this value for future calculation. Active sheet means currently active worksheet, where the cell is located. On the same way we can read all other input parameters that we may be have in our worksheet.

Entering Formulas

Before we start entering formulas, we should have defined all variables for input parameters and also variables for results of calculation.

We enter formulas on the same way as we enter formulas in mathematics.

Let's take the formula below as an example

$$M_{cr} = f_{ctm}bh^2/6000$$

and learn how we will type this formula in Excel. We will write the following statement leading with a comment line as seen below.

```
Debug   Run   Tools   Add-Ins   Window   Help

  ▶  ⅱ  ▫  ▨  ▧  ▧  ▧  ✻  ❓  Ln 28, Col 1

CalcBtn                                         ▼

    Private Sub CalcBtn_Click()

    ' Declaring simple variables
    Dim b As Double
    Dim fctm As Double
    Dim h As Double
    Dim Mcr As Double

    ' Declaring arrays
    Dim Mc(6) As Double

    ' Reading input parameters
    b = ActiveSheet.Cells(19, 4)
    h = ActiveSheet.Cells(20, 4)
    fctm = ActiveSheet.Cells(26, 4)

    ' Cracking moment
    Mcr = fctm * b * (h ^ 2) / 6 / 1000

    End Sub
```

As we can see, we can do it exactly the same way as in mathematics. On the same way we can type all other formulas which we may have.

If we want to know more about VBA – the language that Excel speaks we can click Help in Code Editor for more information. But for simple calculations which we may have it is not needed. Anyway if we want to create more complex calculation with a lot of mathematical functions this help is essential.

3

CALCULATION & RESULTS

In this Chapter

- Displaying Results
- Perform Calculation

Displaying Results

How we can display results of calculation?

As we already prepared Excel cells where results can be displayed, it is obvious that we will write the results of calculation just there. But we will not do it manually, we will rather write a code for it. The code will write the results of calculation which are now stored in variables, directly into these cells.

Let's learn how we can do this.

We will write the following sentence, starting with a comment line.

```
Debug   Run   Tools   Add-Ins   Window   Help
  ▶   II   ■   ◪   ◈   ☞   ☜   ✲   ❷   Ln 27, Col 1

CalcBtn                                    ▼    Click

    Private Sub CalcBtn_Click()

    ' Declaring simple variables
    Dim b As Double
    Dim fctm As Double
    Dim h As Double
    Dim Mcr As Double

    ' Declaring arrays
    Dim Mc(6) As Double

    ' Reading input parameters
    b = ActiveSheet.Cells(19, 4)
    h = ActiveSheet.Cells(20, 4)
    fctm = ActiveSheet.Cells(26, 4)

    ' Cracking moment
    Mcr = fctm * b * (h ^ 2) / 6 / 1000

    ' Display results of curvature calculation
    ActiveSheet.Cells(35, 4) = Mcr

    End Sub
```

It means that the value stored in variable M_{cr} will be written in the Excel cell located at row 35 and column 4. So with this command we wrote the value of this variable directly to the Excel worksheet.

On exactly the same way we can display all other results of calculation from variables in the code to the cells in the Excel worksheet.

Perform Calculation

With this, coding is finished and we need just to Run this code to execute calculation. For this purpose we need to go to the Excel worksheet and make sure that Run mode is active.

How we can do this? As we already said, we need to click the Design Mode button so, that the push button cannot be selected.

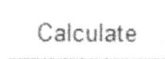

So when we are in Run mode, we need just to click the button and the code is executed and all the results of calculation are displayed in Excel worksheet.

The bottom Line of material presented so far is that we learned how to make a calculation in Microsoft Excel.

But in the following lines we will make one real example of engineering calculation, where you will use the knowledge learned so far and see a real world example.

4

SOFTWARE SAMPLE

In this Chapter

- An introduction to Engineering Calculation
- Cells for input and output parameters
- Entering code editor
- Creating variables and reading data
- Entering formulas
- Entering the For Next loop
- Entering the If statements
- Displaying results and perform calculation

An introduction to Engineering Calculation

In this chapter I will present you the formulas and parameters which we will use in our engineering calculation. We are calculating here deflections of reinforced concrete slab which is a typical structural engineering task. This is just an example of calculation, but you can use the same principles presented here to perform any custom calculation you wish to.

First we have a few calculation formulas for the properties of cracked and un-cracked concrete cross section.

Cracking moment

$$M_{cr} = \frac{f_{ctm} b h^2}{6}$$

Total shrinkage strain

$$\varepsilon_{cs\infty} = \varepsilon_{cd\infty} + \varepsilon_{ca\infty}$$

Modul of elasticity

$$E_{c.eff} = \frac{E_{cm}}{1 + \varphi_k}$$

Ratio of moduluses

$$\alpha_E = \frac{E_s}{E_{c.eff}}$$

Reinforcement ratios

$$\rho = \frac{A_{s1}}{bd} \quad \rho_2 = \frac{A_{s2}}{bd}$$

Virtual moments

$$M_0 = M_6 = 0 \quad M_1 = M_5 = L/12$$
$$M_2 = M_4 = L/6 \quad M_3 = L/4$$

Properties of the uncracked section

$$A_1 = \frac{0.5h^2}{d} + \alpha_E \rho d + \alpha_E \rho_2 d_2 \quad B_1 = \frac{h}{d} + \alpha_E \rho + \alpha_E \rho_2 \quad x_1 = \frac{A_1}{B_1}$$

$$I_1 = \frac{bh^3}{12} + bh(0.5h - x_1)^2 + \alpha_E * A_{s1} * (d - x_1)^2 + \alpha_E * A_{s2}(x_1 - d_2)^2$$

$$S_1 = A_{s1}(d - x_1) - A_{s2}(x_1 - d_2)$$

Properties of the cracked section

$$A_2 = \left(\alpha_E^2 (\rho + \rho_2)^2 + 2\alpha_E \left(\rho + \rho_2 \frac{d_2}{d} \right) \right)^{0.5}$$

$$B_2 = A_2 - \alpha_E (\rho + \rho_2) \quad x_2 = B_2 d$$

$$I_2 = \frac{b}{3} x_2^3 + \alpha_E * A_{s1} * (d - x_2)^2 + \alpha_E * A_{s2}(x_2 - d_2)^2$$

$$S_2 = A_{s1}(d - x_2) - A_{s2}(x_2 - d_2)$$

Flexural curvature calculation

$$i = 0 \rightarrow 6$$

$$r_{1,M,i} = \frac{M_{Ed}^i}{E_{c,eff} I_1} \qquad r_{2,M,i} = \frac{M_{Ed}^i}{E_{c,eff} I_2}$$

$$r_{1,CS,i} = M_{Ed}^i \, \varepsilon_{cs\infty} \alpha_E \frac{S_1}{I_1} \qquad r_{2,CS,i} = M_{Ed}^i \, \varepsilon_{cs\infty} \alpha_E \frac{S_2}{I_2}$$

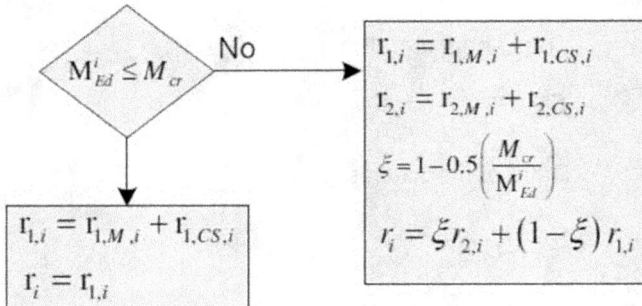

$$M_{Ed}^i \leq M_{cr} \quad \text{No} \rightarrow$$

$$r_{1,i} = r_{1,M,i} + r_{1,CS,i}$$

$$r_{2,i} = r_{2,M,i} + r_{2,CS,i}$$

$$\xi = 1 - 0.5 \left(\frac{M_{cr}}{M_{Ed}^i} \right)$$

$$r_i = \xi r_{2,i} + (1 - \xi) r_{1,i}$$

$$r_{1,i} = r_{1,M,i} + r_{1,CS,i}$$

$$r_i = r_{1,i}$$

We need to perform a calculation in all six points of the moment diagram.

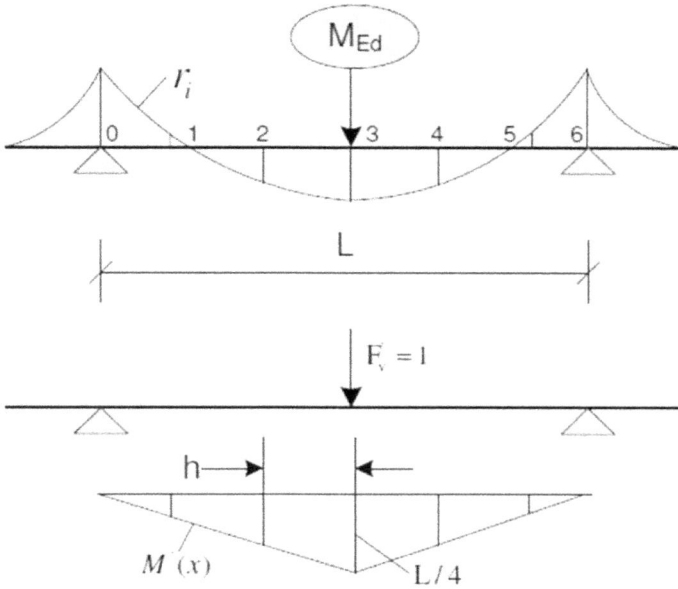

So that we can finally calculate time dependent deflection as an integral, which is approximated with this simple formula as seen below.

$$f_{act} = \int_{0}^{L} r \, M'(x)dx = \int_{0}^{L} f(x)dx$$

$$f_{act} = \frac{h}{3}\left[f_{o} + 4f_{1} + 2f_{2} + 4f_{3} + 2f_{4} + 4f_{5} + f_{6}\right]$$

As each numerical calculation has first input parameters and as a result of calculation we have output parameters or so called variables, we will present them just here for our calculation.

A_{s2}

d h

A_{s1}

b

d_2

x

M_{Ed}

r_i

0 1 2 3 4 5 6

L_{eff}

h

0 1 2 3 4 5 6

$M(x)$ $L/4$

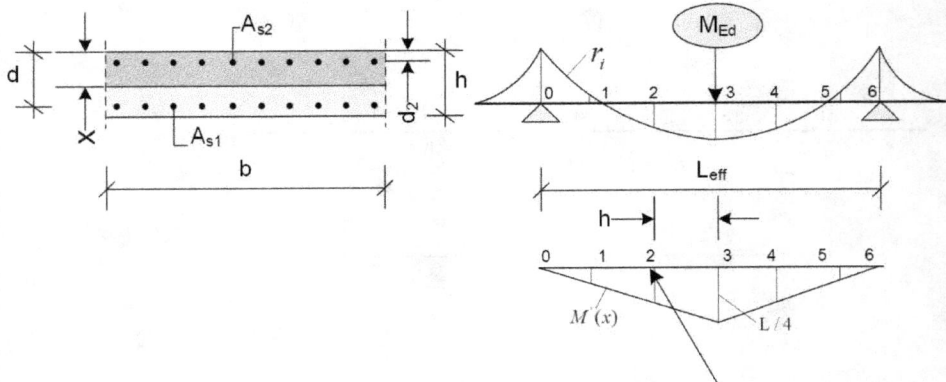

array variables means values calculated in all 6 points shown above, but each value is accessed by an index i=0 to 6!

If we look below, we can see that we have here all **input parameters** which we will need to perform our calculation, they need to be defined before we will start calculating formulas.

b, h width and height of the cross section

d effective depth to tension reinforcement

d_2 effective depth to compression reinforcement

f_{ctm} tensile strength of concrete

$\varepsilon_{cd\infty}$ drying shrinkage strain of concrete

$\varepsilon_{ca\infty}$ autogenous shrinkage strain of concrete

φ_k creep coefficient of concrete

E_{cm} modulus of elasticity of concrete

E_s modulus of elasticity of reinforcement

A_{s1} area of lower reinforcement

A_{s2} area of upper reinforcement

L_{eff} effective span length

M_{Ed} bending moment

Below we have output parameters where we store the results of calculation. This parameters are defined as simple variables.

M_{cr} cracking moment

$\varepsilon_{cs\infty}$ total shrinkage strain of concrete

$E_{c,eff}$ effective modulus of elasticity of concrete

α_E ratio of moduluses

ρ, ρ_2 reinforcement ratios

f_{act} final deflection

Properties of the uncracked section
A_1, B_1, x_1, I_1, S_1

Properties of the cracked section
A_2, B_2, x_2, I_2, S_2

Below we have output parameters for results which are presented in a table as array variables.

$i = 0 \rightarrow 6$

M_i' virtual moment

$r_{1,M}^i$ flexural curvature for uncracked concrete

$r_{2,M}^i$ flexural curvature for cracked concrete

$r_{1,CS}^i$ curvature due to shrinkage strain for uncracked concrete

$r_{2,CS}^i$ curvature due to shrinkage strain for cracked concrete

ς^i curvature reduction coefficient

r^i final curvature

What is array we already learned in this book, so we will not repeat it once more! We will just emphasize that all array variables are calculated in all 6 points of the moment diagram and for final deflection calculation the integral of them is solved.

Cells for input and output parameters

As we already learned, we need for every calculation first to prepare cells for input and output parameters. How to create these cells we already learned in this book, so we will not type here all input parameters as it is exactly the same for all parameters, we will just show the final result.

Input parameters			
b=	100,00 cm	Width of the cross section	Calculate
h=	15,00 cm	Height of the cross section	
d=	13,00 cm	Effective depth to tension reinforcement	
d_2=	2,00 m	Depth to compression reinforcement	
L_{eff}=	5,00 m	Effective span length	
A_s=	5,24 cm²	Area of tension reinforcement	
A_{s2}=	0,00 cm²	Area of compression reinforcement	
f_{ctm}=	2,60 N/mm²	Mean value of axial tensile strength of concrete	
E_{cm}=	31.000 N/mm²	Secant modulus of elasticity of concrete	
E_s=	200.000 N/mm²	Design value of modulus of elasticity of reinforcing steel	
φ_∞ =	2,60	Final value of creep coefficient	
$\varepsilon_{cd\infty}$ =	0,50 ‰	Final value of drying shrinkage strain	
$\varepsilon_{ca\infty}$ =	0,10 ‰	Final value of autogenous shrinkage strain	

Curvature calculation		
M_{cr}=	kNm	Cracking moment
$E_{c,eff}$=	N/mm²	Effective modulus of elasticity of concrete
$\varepsilon_{cs\infty}$ =	‰	Final total value of shrinkage strain
α_E =		Modular ratio

T_i	M_{Ed} [kNm]	M' [kNm]	$(1/r)_{I,M}$ [1/m]	$(1/r)_{I,CS}$ [1/m]	$(1/r)_{II,M}$ [1/m]	$(1/r)_{II,CS}$ [1/m]	ξ	$(1/r)$ [1/m]
0	0,00							
1	6,50							
2	8,50							
3	9,80							
4	8,50							
5	6,50							
6	0,00							

Deflection calculation		
f_{act}=	mm	Final deflection

As we can see, we have here created cells for all input parameters – we can see here width and height of cross section, effective depth, length, reinforcement data and properties of the concrete.

Part of input parameters are displayed in a table – we can see here data for bending moment in all 6 points of the moment diagram.

For easier work we have text color of input cells in blue color.

For output parameters we need first to prepare cells for results of calculation of curvature ductility. We have four cells which are created on the way we already learned, so we will not type each cell at this time, but we will just show final result of our entry – we can see these cells including with their units and descriptions.

Part of our results are displayed in a table, so we need to prepare table header and format the numerical values of the cells. As we already learned, we can see cells created and formatted according to our requirements.

Finally we need to prepare also the cell for final deflection calculation. We can see this cell here, including header, units and description.

Now we need just to write the code to perform the calculation, which will be described in detail in the following lines.

Entering Code Editor

First we need to access the code editor. For this task, we need to have Developer Tab visible in the Excel Ribbon. How to do this we already learned in this book and we will not repeat it here again.

Now we need to select a Developer Tab and click Design mode button here, so that we are in design mode. It enables us to enter code editor, when we double click on the push button what we already learned in this book.

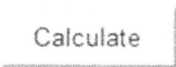

Calculate

You can check if you are in design mode, if the push button can be selected, otherwise you need to click Design Mode button once more.

So when we are in Design Mode we just double click the Push Button and we enter the Code Editor.

File   Edit   View   Insert   Format   Debug   Run   Tools   Add-Ins   Window   Help

Project - VBAProject

VBAProject (Eng-Calc.xlsm)
Microsoft Excel Objects
Sheet25 (Def-calc)
Sheet26 (1)
Sheet27 (0)
ThisWorkbook
Modules

CalcBtn   Click

```vba
Private Sub CalcBtn_Click()

' Declaring simple variables
Dim b As Double
Dim fctm As Double
Dim h As Double
Dim Mcr As Double

' Declaring arrays
Dim Mc(6) As Double

' Reading input parameters
b = ActiveSheet.Cells(19, 4)
h = ActiveSheet.Cells(20, 4)
fctm = ActiveSheet.Cells(26, 4)

' Cracking moment
Mcr = fctm * b * (h ^ 2) / 6 / 1000

' Display results of curvature calculation
ActiveSheet.Cells(35, 4) = Mcr

End Sub
```

Watches

Expression	Value	Type	Context

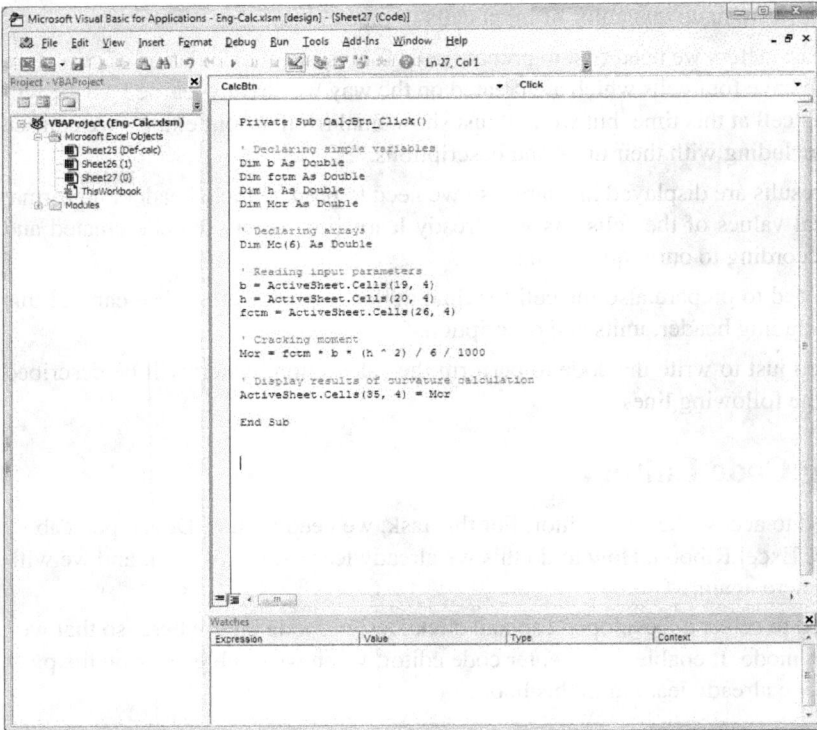

We can see here the code we already typed in this book. Now we will implement it so, that our calculation will be complete.

Creating variables and reading data

First we will create simple variables the same way as we already did for a few variables. For all parameters described in this chapter we need to create variables. We showed here declaration of all simple variables in our calculation.

CalcBtn   ▼   Clic

```
Private Sub CalcBtn_Click()

' Declaring simple variables
Dim b, h, d, d2 As Double
Dim Leff As Double
Dim fctm, Ecm, Es As Double
Dim Eceff As Double
Dim As1, As2 As Double
Dim fik, epscdk, epscak, epscs As Double
Dim Mcr, alphae As Double
Dim ro, ro2 As Double
Dim A1, B1, k1, kx1 As Double
Dim x1, I1, S1 As Double
Dim A2, B2, k2, kx2 As Double
Dim x2, I2, S2 As Double
Dim MEd As Double
Dim fact As Double
```

If the result of calculation is not stored just in one variable but instead in a column, we need to create arrays. For all arrays described in this chapter we need to create array variables. In our example we will create arrays which will hold 6 elements. Below you can see all array variables which we created for our calculation.

CalcBtn   ▼

```
Dim Mcr, alphae As Double
Dim ro, ro2 As Double
Dim A1, B1, k1, kx1 As Double
Dim x1, I1, S1 As Double
Dim A2, B2, k2, kx2 As Double
Dim x2, I2, S2 As Double
Dim MEd As Double
Dim fact As Double

' Declaring arrays
Dim Mc(6) As Double
Dim r1M(6) As Double
Dim r2M(6) As Double
Dim r1CS(6) As Double
Dim r2CS(6) As Double
Dim r1(6) As Double
Dim r2(6) As Double
Dim r(6) As Double
Dim ksi(6) As Double
Dim fx(6) As Double
```

Before we can start our calculation, input values in cells should be stored in input parameters. This can be done with the following code which we already explained in this book. We can see below all the values from input cells stored in previously defined variables.

Debug   Run   Tools   Add-Ins   Window   Help

▷   Ln 45, Col 34

CalcBtn   ▾

```
Dim r1(6) As Double
Dim r2(6) As Double
Dim r(6) As Double
Dim ksi(6) As Double
Dim fx(6) As Double

' Reading input parameters
b = ActiveSheet.Cells(19, 4)
h = ActiveSheet.Cells(20, 4)
d = ActiveSheet.Cells(21, 4)
d2 = ActiveSheet.Cells(22, 4)
Leff = ActiveSheet.Cells(23, 4)
As1 = ActiveSheet.Cells(24, 4)
As2 = ActiveSheet.Cells(25, 4)
fctm = ActiveSheet.Cells(26, 4)
Ecm = ActiveSheet.Cells(27, 4)
Es = ActiveSheet.Cells(28, 4)
fik = ActiveSheet.Cells(29, 4)
epscdk = ActiveSheet.Cells(30, 4)
epscak = ActiveSheet.Cells(31, 4)
```

Entering Formulas

Before we start entering formulas, we should have defined all variables for input parameters and also variables for results of calculation.

We enter formulas on the same way as we enter formulas in mathematics. We already learned how we enter formulas into Excel in this book.

Each of the formulas defined in this chapter need to be written in code editor. We will not type them manually at this time, rather we will just present them entered in the code editor.

CalcBtn   ▼   Click

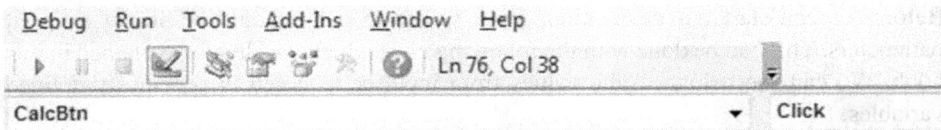

```
epscdk = ActiveSheet.Cells(30, 4)
epscak = ActiveSheet.Cells(31, 4)

' Cracking moment
Mcr = fctm * b * (h ^ 2) / 6 / 1000

' Effective modul of elasticity of concrete
Eceff = Ecm / (1 + fik)

' Total shrinkage strain
epscs = epscdk + epscak

' Ratio of moduls of elasticity of concrete and steel
alphae = Es / Eceff

' Reinforcement ratios
ro = As1 / (b * d)
ro2 = As2 / (b * d)

' Properties of the uncracked section
A1 = 0.5 * (h ^ 2) / d + alphae * ro * d + alphae * ro2 * d2
B1 = h / d + alphae * ro + alphae * ro2
x1 = A1 / B1
I1 = b * (h ^ 3) / 12 + b * h * ((0.5 * h - x1) ^ 2)
I1 = I1 + alphae * As1 * ((d - x1) ^ 2) + alphae * As2 * ((x1
S1 = As1 * (d - x1) - As2 * (x1 - d2)

' Properties of the cracked section
A2 = ((alphae ^ 2) * ((ro + ro2) ^ 2) + 2 * alphae * (ro + ro:
B2 = A2 - alphae * (ro + ro2)
x2 = B2 * d
I2 = b / 3 * (x2 ^ 3) + alphae * As1 * ((d - x2) ^ 2) + alpha
S2 = As1 * (d - x2) - As2 * (x2 - d2)|
```

As already said if we want to know more about VBA – about the language that Excel speaks we can click Help in Code Editor for more information.

But for simple calculations as we have here it is not needed.

Entering the For-Next Loop

As we want to calculate values for arrays, we need to repeat the calculation for each value of the index, which has in our example value from 0 to 6. For this purpose we have to write in the code editor the following text.

```
For i = 0 To 6

Next i
```

Whatever we insert between those lines will be repeated for the value of variable i from 0 to 6 with increment 1.

So in our example we will insert here the following formulas, which we need in our calculation.

```
' Flexural curvature
For i = 0 To 6
  If i = 0 Then
    Mc(i) = 0
  ElseIf i = 1 Then
    Mc(i) = Leff / 12
  ElseIf i = 2 Then
    Mc(i) = Leff / 6
  ElseIf i = 3 Then
    Mc(i) = Leff / 4
  ElseIf i = 4 Then
    Mc(i) = Leff / 6
  ElseIf i = 5 Then
    Mc(i) = Leff / 12
  ElseIf i = 6 Then
    Mc(i) = 0
  End If
  MEd = ActiveSheet.Cells(42 + i, 4)
  r1M(i) = MEd / (Eceff * I1) * 100000
  r2M(i) = MEd / (Eceff * I2) * 100000
  r1CS(i) = MEd * epscs * alphae * S1 / I1 / 10
  r2CS(i) = MEd * epscs * alphae * S2 / I2 / 10
  If Abs(MEd) < Mcr Then
    r1(i) = r1M(i) + r1CS(i)
    r(i) = r1(i)
  Else
    r1(i) = r1M(i) + r1CS(i)
    r2(i) = r2M(i) + r2CS(i)
    ksi(i) = 1 - 0.5 * Mcr / Abs(MEd)
    r(i) = r2(i) * ksi(i) + r1(i) * (1 - ksi(i))
  End If
  fx(i) = r(i) * Mc(i)
Next i
```

That way we will calculate array values in all 6 points of moment diagram.

At the end we will also enter the formulas for final deflection calculation as we can see below.

```
' Deflection calculation
fact = fx(0) + 4 * fx(1) + 2 * fx(2) + 4 * fx(3) +
fact = fact * (Leff / 6) / 3 * 1000
```

Entering the If statements

In our calculation we haven't entered just simple formulas. We entered also the **If statement** which I would like to explain here.

Let's look once again at our calculation. As we can see, different formulas are used, depending of the fulfilment of one condition. This is achieved in the code with the following statements.

```
If Condition Then
   Statement1
ElseIf Condition1 Then
   Statement2
ElseIf Condition2 Then
   Statement3
Else
   Statement4
End If
```

In our calculation this If structure is used, so our formulas are calculated using the conditioning of the calculation requirements we have.

Displaying results and perform calculation

How we can display results of calculation?

As we already prepared Excel cells where results can be displayed, it is obvious that we will write the results of calculation just there.

But we will not do it manually, we will write the code for it. The code will write the results of calculation which are now stored in variables, directly into these cells.

As we already learned the basics of this code in this book, we will just present them here.

```
' Display results of curvature calculation
ActiveSheet.Cells(35, 4) = Mcr
ActiveSheet.Cells(36, 4) = Eceff
ActiveSheet.Cells(37, 4) = epscs
ActiveSheet.Cells(38, 4) = alphae

' Fill curvature table
For i = 0 To 6
  ActiveSheet.Cells(42 + i, 5) = Mc(i)
  ActiveSheet.Cells(42 + i, 6) = r1M(i)
  ActiveSheet.Cells(42 + i, 7) = r1CS(i)
  ActiveSheet.Cells(42 + i, 8) = r2M(i)
  ActiveSheet.Cells(42 + i, 9) = r2CS(i)
  ActiveSheet.Cells(42 + i, 10) = ksi(i)
  ActiveSheet.Cells(42 + i, 11) = r(i)
Next i

' Display deflection
ActiveSheet.Cells(52, 4) = fact
```

With this code we will just write the values stored in variables to the cells prepared in Excel where results of calculation are displayed.

So now we can close and save the code editor and switch to the Excel worksheet where we have our data displayed.

With this, coding is finished and we need just to run this code to execute calculation.

As we already learned about Run and Design mode of Excel, we will just remind you here, that you need to be in Run mode and click the Push button.

When you will do this, results of the calculation will be at once displayed in worksheet as we can see on the next page.

Curvature calculation

$M_{cr} =$ 9,75 kNm — Cracking moment
$E_{c,eff} =$ 8.611 N/mm^2 — Effective modulus of elasticity of concrete
$\varepsilon_{cs\infty} =$ 0,60 ‰ — Final total value of shrinkage strain
$\alpha_E =$ 23,23 — Modular ratio

T_i	M_{Ed} [kNm]	M' [kNm]	$(1/r)_{I,M}$ [1/m]	$(1/r)_{I,CS}$ [1/m]	$(1/r)_{II,M}$ [1/m]	$(1/r)_{II,CS}$ [1/m]	ξ	$(1/r)$ [1/m]
0	0,00	0,00	0,0000	0,0000	0,0000	0,0000	0,00	0,0000
1	6,50	0,42	0,0024	0,0077	0,0064	0,0340	0,00	0,0101
2	8,50	0,83	0,0031	0,0100	0,0083	0,0444	0,00	0,0131
3	9,80	1,25	0,0036	0,0115	0,0096	0,0512	0,50	0,0381
4	8,50	0,83	0,0031	0,0100	0,0083	0,0444	0,00	0,0131
5	6,50	0,42	0,0024	0,0077	0,0064	0,0340	0,00	0,0101
6	0,00	0,00	0,0000	0,0000	0,0000	0,0000	0,00	0,0000

Deflection calculation

$f_{act} =$ 74,39 mm — Final deflection

5

IN CLOSING

So this is it. We created a simple software for our engineering calculation. I hope I convinced you that you do not need to be a software developer to write your own customized engineering calculation in minutes.

If you want to know more about VBA – the language in which Excel speaks you can click Help in Code Editor for more information. But for simple calculations which we may have it is not needed. Anyway if you want to create more complex calculation with a lot of mathematical functions this help is essential.

I provide also online course on Engineering Calculations using Microsoft Excel. You will be offered full video presentation, where I can show you live in Microsoft Excel how you can do it.

You can get more information about the course at the following address:
https://www.udemy.com/engineering-calculations-using-microsoft-excel/#/

www.ingramcontent.com/pod-product-compliance
Lightning Source LLC
Chambersburg PA
CBHW061840220326
41599CB00027B/5358